Jupiter

Tim Goss

Heinemann Library
Chicago, Illinois

Layout by Roslyn Broder
Illustrations by Calvin J. Hamilton
Printed in Hong Kong

07 06 05 04 03
10 9 8 7 6 5 4 3 2 1

Library of Congress Cataloging-in-Publication Data
Goss, Tim, 1958-
  Jupiter / by Tim Goss.
    v. cm. -- (The universe)
Includes bibliographical references and index.
Contents: Where in the sky is Jupiter? -- What's special about Jupiter?
-- Does Jupiter have moons? -- How's the weather on Jupiter? -- What
would I see if I went to Jupiter? -- What's inside Jupiter?
-- Could I ever go to Jupiter?
  ISBN 1-58810-911-9 (HC), I-4034-0612-X (Pbk)
  1. Jupiter (Planet)--Juvenile literature. [1. Jupiter (Planet)] I.
Title. II. Series.
  QB661 .G67 2002
  523.45--dc21
                                2002000815

Acknowledgments
The author and publisher are grateful to the following for permission to reproduce copyright
material: pp. 4, 5, 9, 12, 15, 18, 20, 25, 27, 28 NASA/JPL/Caltech; p. 6 D. Van Ravenswaay/Photo
Researchers, Inc.; pp. 7, 8, 13, 19, 21, 29 NASA/JPL/University of Arizona; p. 10 NASA/JPL/Cornell
University; p. 11 Bettmann/Corbis; pp. 14, 16 NASA/JPL/DLR (German Aerospace Center); p. 17
NASA/JPL/U.S. Geological Survey; p. 22 NASA/JPL/California Institute of Technology; p. 23 NASA/ESA,
John Clarke/University of Michigan, NASA and the Hubble Heritage Team (STScI/AURA); p. 24
Courtesy of Calvin J. Hamilton/www.solarviews.com; p. 26 NASA

Cover photograph by NASA/JPL/Caltech

The publisher would like to thank Geza Gyuk and Diana Challis of the Adler Planetarium for their
comments in the preparation of this book.

Every effort has been made to contact copyright holders of any material reproduced in this book.
Any omissions will be rectified in subsequent printings if notice is given to the publisher.

Some words are shown in bold, **like this.** You can find out what
they mean by looking in the glossary.

Contents

Where in the Sky Is Jupiter?

The mark for which the planet Jupiter is best known is the large spot that you see about halfway down on the striped planet.

When Jupiter appears in the night sky, you can see it without a **telescope**. It looks like a bright silver star.

Jupiter is the fifth planet from the Sun in our **solar system**. It is also the second brightest planet. Venus is the only planet that is brighter.

King of the planets

Jupiter is the largest planet in our solar system. Its diameter, the distance right through the middle of the planet from one side to the other, is 11 times as large as Earth's. That means you could put 11 Earths side by side inside Jupiter. If you stuffed as many Earths as possible inside of Jupiter, 1,318 Earths would fit.

The king's colors

If you view Jupiter through a telescope, you can see that it has alternating stripes of cream and brown. The cream stripes are called **zones.** The brown stripes are called **belts.**

Different **chemicals** in each layer make the different colors. Most of Jupiter is made up of clouds and gas layers. It is one of the planets called the Gas Giants. The others are Saturn, Uranus, and Neptune.

When an image of Jupiter is placed with images of the other planets, its larger size is easy to recognize. Here, Jupiter is the fifth planet from the top.

The solar system

A solar system is a group of objects in space that move in orbits around a central star. The objects can be planets, moons, **comets, meteors,** and **asteroids.** The Sun is the central star in our solar system. **Gravity** from the Sun pulls on all of the objects in our solar system. This keeps Jupiter and the other objects in our solar system from flying off into space.

How long is a Jupiter year?

One **year** is the time it takes for a planet to make one circle, or **revolution,** around the Sun. On Earth, a year is 365 days long. It takes Jupiter almost twelve Earth years to make one revolution. As it **orbits** the Sun, Jupiter travels at about 29,000 miles (47,000 kilometers) per hour. Earth orbits the Sun at a faster speed of about 66,600 miles (107,000 kilometers) per hour. When you combine Jupiter's slower orbital speed with its greater distance from the Sun, you can understand why Jupiter's year is so long.

Jupiter can be seen from Earth without a ***telescope*** *even though it is usually about 480 million miles (772 million kilometers) away.*

The difference in the revolutions of Earth and Jupiter is what causes Jupiter's different positions in the sky when you look at it from Earth. Earth keeps passing

Jupiter as the planets orbit the Sun because Earth's revolution is much faster than Jupiter's. As Earth is about to catch up to and pass Jupiter, Jupiter appears to be in front of Earth. After Earth passes Jupiter, Jupiter appears to be behind Earth.

Jupiter days are very short

A **day** is the time it takes for a planet to spin around once on its **axis,** the imaginary line that runs from a planet's north pole to its south pole. One Earth day is 24 hours long. A day on Jupiter lasts a little less than ten hours. Jupiter spins faster on its axis than any other planet.

How did Jupiter get its name?

Astronomers named Jupiter after the king of the Roman gods. This god was the ruler of everything in the sky. He was believed to be the god who brought light to the world. Paintings of the god Jupiter often show him with a lightning bolt in his hand.

What's Special About Jupiter?

Everything weighs more on Jupiter

The **gravity** on Jupiter is much stronger than on Earth. It is about two and a half times stronger. If you weigh 80 pounds (36 kilograms) on Earth, you would weigh 200 pounds (90 kilograms) on Jupiter. That's because the force holding you to the **surface** of the planet is pulling you down much harder.

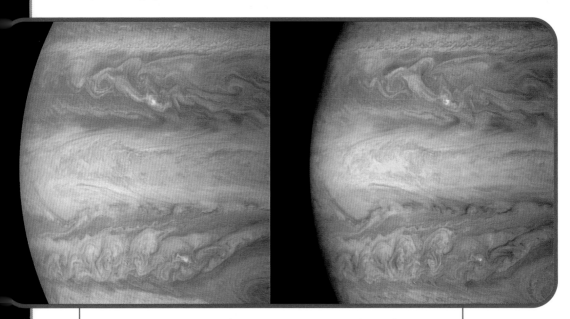

The image on the left shows the true color of Jupiter. Color filters were used in the image on the right to help scientists study the planet. The darkest blue areas are called hot spots.

Jupiter makes its own heat

On Earth, the Sun provides nearly all of the warmth that we feel. The Sun's heat energy warms the ground and the oceans. Earth cannot give off any more energy than it receives. Jupiter, however, gives off more heat than it gets from the Sun. How does this happen?

Jupiter's **core** is in the very center of the planet. The temperature there is about 54,000°F (30,000°C), and it used to be even hotter! Heat energy from the core rises to the surface. Where did this heat come from and why has it not run out? When each planet first formed from bits of dust, rocks, and ice, a lot of heat was created when gravity squeezed these objects together. This heat of formation has been escaping slowly from the planets over millions of years. Small planets, like Earth, lose their heat quickly. Large planets, like Jupiter, have much more heat to begin with and lose it very slowly.

Jupiter has rings

Jupiter has a narrow set of **rings** that you can see with a **telescope.** Information from the *Pioneer 10* and *11* missions in 1973 and 1974 led scientists to think there might be rings around Jupiter. No one saw the rings until the *Voyager I* mission sent back images in 1979. The planet Saturn also has rings, but Jupiter's rings are smaller and darker. Jupiter's rings seem to be made of dust and rock particles.

This image gives a close-up look at Jupiter's ring system.

Jupiter is a giant magnet

Jupiter is one of six planets that have **magnetic fields** bigger than the planets themselves. The other planets are Earth, Mercury, Saturn, Uranus, and Neptune. A magnetic field cannot be seen. It fills the areas around a magnet. Inside a magnetic field, the motion of electric charges and other magnets are affected.

Gossamer Rings
Main Ring
Halo
Amalthea
Adrastea
Metis
Thebe

*Jupiter's large magnetic field reaches beyond the planet's system of **rings** and **moons**.*

The magnetic field around Jupiter is huge. It stretches away from the planet at least 4 million miles (6.4 million kilometers) in every direction. That is bigger than the Sun!

Particles flowing out from the Sun, called the solar wind, sweep Jupiter's magnetic field into a long "tail." Jupiter's magnetic tail stretches more than 600 million miles (966 million kilometers). It goes beyond the **orbit** of Saturn!

Does Jupiter Have Moons?

Galileo Galilei's discovery

On January 6, 1610, an **astronomer** named Galileo Galilei saw what appeared to be three stars lined up with Jupiter. When he looked the next night, the positions of the stars had changed. The positions continued to change every night. Galileo realized that the objects he was looking at were moons. They were orbiting Jupiter just like Earth's Moon goes around Earth. Later, Galileo saw that there were actually four moons orbiting Jupiter.

Galileo's discovery of Jupiter's moons was one of the most important discoveries ever made. The old ideas about the universe were that the Earth stayed still and everything else moved around it. But the moons Galileo discovered moved around Jupiter, not around Earth. His discovery supported the ideas of scientists Nicolaus Copernicus and Johannes Kepler. They thought Earth and the other planets moved around the Sun.

The Italian astronomer Galileo was born in 1564 and died in 1642. During his lifetime, he made many important astronomical discoveries.

Jupiter's four closest moons are much closer to Jupiter than Earth's Moon is to Earth. This is an artist's impression of what Jupiter and its closest moons would look like from near the moon Callisto (lower right). The other three moons, top to bottom, are Io, Europa, and Ganymede.

Astronomers have found sixteen moons

Jupiter has four tiny **moons** that are very close to it. Three of these tiny moons were discovered in 1979. The fourth one was discovered more than 100 years ago in 1892. Next are the four huge moons discovered by Galileo in 1610, known as the Galilean moons. The other eight, farther from Jupiter, are very small. A moon called Sinope is the farthest moon from the planet. It is about 15 million miles (24 million kilometers) from Jupiter. To travel this distance in space, you would have to go to Earth's Moon and back almost 32 times!

There may be more moons. **Astronomers** at the University of Hawaii have discovered eleven small objects that may be more moons **orbiting** Jupiter.

The Galilean moon Io

Io is about the size of Earth's Moon. Io is mostly flat, but there are mountains on it that are almost six miles (about 10 kilometers) tall. In 1979, *Voyager 1* discovered active **volcanoes** on Io. As far as scientists know, only Earth and Io have active volcanoes. Io's volcanoes are more active and hotter than Earth's. The *Voyager* missions discovered more than sixteen volcanoes erupting on Io.

Most of Io's **craters** were caused by volcanic actions. The largest volcano on Io, called Pele, throws out material into an area the size of the state of Arizona. Io's **surface** is spotted with red, yellow, white, and orange-black. The colors are from the **sulfur** in Io's top layer, the **crust.**

Io has more active volcanoes than any other object in the solar system.

The Galilean moon Europa

Europa is almost as big as Earth's Moon. There are very few **craters** on Europa. It is the smoothest object in the **solar system.** Because there are not very many craters, scientists think that the **surface** of Europa is very young compared to that of the other **moons.** The surface might only be a few million years old. Europa's icy surface reflects a lot of sunlight. It reflects about five times more light than Earth's Moon reflects.

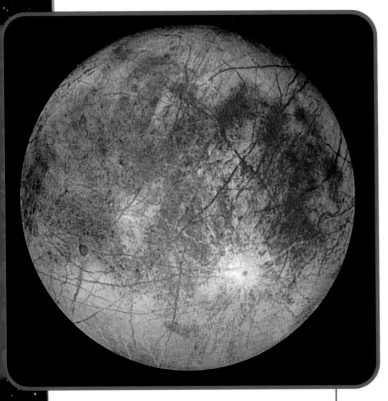

The **crust** of Europa is made of water ice and is probably about 60 miles (100 kilometers) thick, though no one knows for certain. Some large areas on Europa have lots of icebergs. Below Europa's surface, temperatures may be warm enough for an ocean of water.

Images of Europa show thousands of cracks and ridges crisscrossing one another over and over again. Some of these cracks are thousands of miles long.

The Galilean moon Ganymede

Ganymede is the largest moon in our solar system. It is more than one and a half times bigger than Earth's Moon. It is also bigger than two planets—Mercury and Pluto. Part of Ganymede's surface is covered by dark areas with lots of craters. The rest is lighter and has grooves, which are long lines that cut into the moon's surface. Some grooves are thousands of miles long.

Ganymede has its own **magnetic field.** This makes scientists think that it has metals in its **core.**

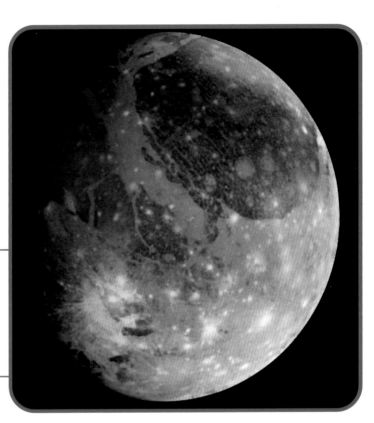

The Galilean moon Callisto

Callisto is the Galilean moon farthest from Jupiter. It is the third largest moon in our solar system. Only Ganymede and Saturn's moon Titan are bigger. Callisto's surface is covered with cracked ice and craters. Some of the craters are very large. Callisto has more craters than any other object in our solar system.

Callisto's **craters** are called impact craters. They formed when **meteorites, comets,** and **asteroids** crashed into its **surface.** Callisto's many craters have made scientists think that the **moon's** surface is very old. In fact, they think Callisto's surface has not been remade for billions of years. If **molten** rock or ice had covered the surface since then, most of the craters would have been wiped away. Only a very old surface would have had time to collide with so many objects.

The Galileo *spacecraft discovered that like the moon Europa, Callisto might have an ocean under its icy* crust. *Scientists have also learned from the* Galileo *that* oxygen *may be inside the ice and rocks on Callisto.*

How's the Weather on Jupiter?

Since Jupiter is so far away from the Sun, it is always cold. At the top of Jupiter's cloud layer, the temperatures get as cold as –235°F (–150°C). It is also very windy. The fastest winds, near the planet's **equator,** can reach speeds of up to 340 miles (550 kilometers) per hour. There are also thunderstorms.

Cloudy

Jupiter's **atmosphere** is made up of mostly **hydrogen** and **helium**. There are three cloud layers within the atmosphere, which are about 30 miles (50 kilometers) thick from top to bottom. The top layer of clouds is made of **ammonia** ice. The middle layer has crystals made of a combination of ammonia, hydrogen, and **sulfur.** The bottom layer contains water ice and maybe even liquid water.

The **chemicals** in Jupiter's cloud layers react with the tiny bit of **carbon** in the planet's atmosphere. This creates new chemical **compounds** in different colors. These are the reds, browns, and blues we see in Jupiter's clouds.

Some areas of the clouds look dark and others appear lighter. The dark areas are called **belts** and the light areas are called **zones.** These form the colored bands we see around the planet. Belts and zones are found in different places at different times. They also change size.

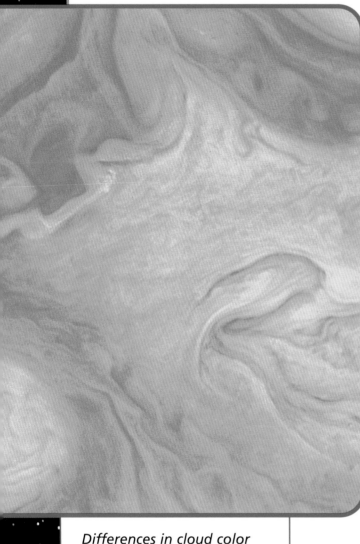

Differences in cloud color may be based on the heights of the clouds. Scientists do not know for sure.

Windy

Winds on Jupiter are formed in a different way than those on Earth. The winds on Earth are caused by differences in temperatures. Hot air rises and cold air sinks. The greater the difference in temperature, the stronger the wind when the two air masses run into each other. The Sun warms Earth's **atmosphere.** It is often about 100°F (38°C) warmer at the **equator** than at the north or south pole.

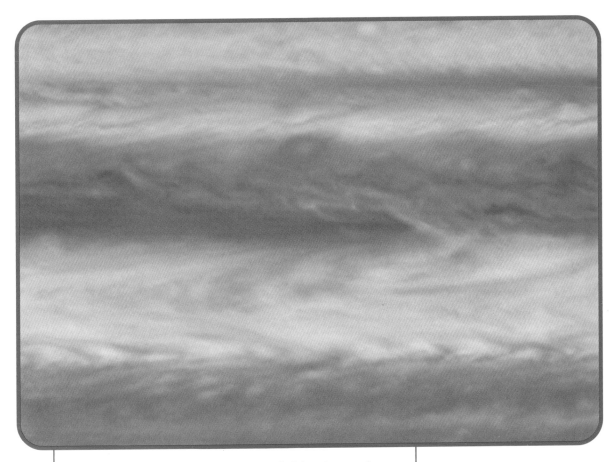

The blue area in the center of this picture is a warmer, cloud-free area called a hot spot.

On Jupiter, the temperatures at the north and south poles and the equator are almost the same. Because it is so far away, the Sun does not have as much effect on Jupiter as it does on Earth. Scientists know that Jupiter's atmosphere, which is below the planet's clouds, is very windy. Heat from Jupiter's **core** flows out into the atmosphere and warms the clouds. This probably creates the wind currents.

Storms and Spots

The storms inside Jupiter's **belts** and **zones** have lasted for many years. One storm called the Great Red Spot has lasted for at least 300 years. The Great Red Spot is an area of dark red and brown clouds that spin in a circle. The clouds complete one circle every six days. The winds blow at about 250 miles (400 kilometers) per hour. The Great Red Spot covers an area bigger than two Earths. It is the most powerful storm in the **solar system.**

The white storm

The *Galileo* space mission returned images of two white storm centers combining into one big storm. This huge storm had an oval shape, like a chicken egg, and it is still active. The only storm bigger than this one is the Great Red Spot.

This image shows a close-up of the Great Red Spot.

What Would I See if I Went to Jupiter?

Color filters make the planet's belts and zones and Great Red Spot really stand out.

To visit Jupiter, you would need a lot of equipment to survive. You would need tanks of air so you could breathe. The **atmosphere** below the clouds is a poisonous mixture of **hydrogen** and **helium** gases, with tiny amounts of **ammonia** and **methane.** Since Jupiter is a Gas Giant, there is no solid ground to land on. You would need a spacecraft that could fly. As you dropped further down toward the **core,** the hydrogen and helium gases would slowly change into a liquid form. Your spacecraft would need to be a submarine, too!

A deep sea

If you swim to a great depth in an ocean on Earth, your ears will "pop" from the **pressure** of the water. The same thing would happen on Jupiter as you moved from the clouds to deep inside the planet.

The weight of all the thousands and thousands of miles of **atmosphere** and liquid **hydrogen** and **helium** would crush you. Only high in the atmosphere, where the clouds are, is the **pressure** similar to the pressure of the air we breathe on Earth.

By the time you went about 13,000 miles (21,000 kilometers) down into Jupiter's sea, the pressure would be more than 3 million times what it is on Earth's **surface.** That is enough to make the liquid hydrogen turn into metal. No one knows much about liquid metallic hydrogen because it cannot exist on Earth. This sea of metallic hydrogen is about 25,000 miles (40,000 kilometers) deep.

This is a view from between cloud layers on Jupiter. False color is added to show the different kinds of clouds.

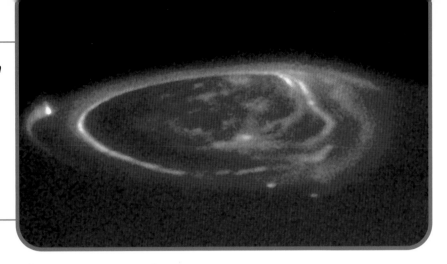

An **aurora** is a reaction between particles in the **magnetic field** of a planet and gases in the planet's atmosphere. Jupiter's auroras create a light show of electric blue.

A super light show on Jupiter

If you had visited Jupiter in July 1994, you would have seen a **comet** crash into the planet. Scientists first saw the comet in March 1993. As they studied it, they learned that in early July 1992, the comet had come close enough to Jupiter to break into 21 pieces. Some of these pieces were over a mile (two kilometers) wide. Scientists were able to predict when pieces of the comet would crash into Jupiter.

In July 1994, the pieces plunged into Jupiter's atmosphere and exploded from the impact. The explosions were so bright that **astronomers** on Earth could see the light reflected from Jupiter's **moons.** The *Galileo* spacecraft took pictures of the collisions.

What makes the magnetic field magnetic?

The high pressure in the sea of liquid metallic hydrogen squeezes the hydrogen **molecules** so tightly that they break apart. It is like trying to stuff a lot of grapes into a box. If you pack them too tightly, the grapes will burst. When hydrogen becomes metallic, electricity flows through it much more easily. Scientists think the electric currents in the liquid metallic hydrogen create Jupiter's magnetic field.

What Is Inside Jupiter?

Looking at the inside of Jupiter, one of the Gas Giants, is different than looking at the inside of one of the Rocky Planets, such as Earth. Below the **atmosphere** on Earth is the solid **surface** of the planet. The **crust** is below the surface. Under the crust is the **mantle,** which is almost completely solid. Finally, there is the **core,** which has both solid parts and **molten** parts.

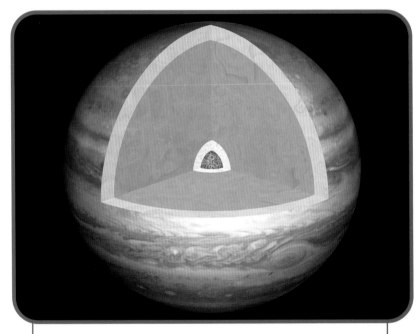

Jupiter, like all planets, has different layers inside it. These layers all blend together, though, inside the Gas Giants.

On Jupiter, there is no surface or crust. The planet's atmosphere of **hydrogen** and **helium** simply gets thicker and denser the farther toward the center you go. The region of liquid metallic hydrogen can be thought of as an inner mantle.

Jupiter's core

Jupiter does have a huge core, and it might be solid. Scientists think the core is made of rock and metal, but no one knows for sure. Jupiter's core is about the size of two Earths. The temperature of the core is about 30,000°F (17,000°C). The slow escape of this heat affects the winds on Jupiter. Some of the heat gets through holes in the clouds, creating areas called hot spots.

The Galileo *probe's heat shield and parachute helped it enter Jupiter's atmosphere.*

We don't know much about what's inside Jupiter

Scientists cannot measure the inside of Jupiter directly. The hot temperatures would melt any **probe** that tried to go into the interior of the planet. The high **pressures** would also crush scientific instruments. A probe dropped by the *Galileo* spacecraft lasted less than an hour.

Could I Ever Go to Jupiter?

To visit Jupiter, you would need a lot of special equipment to survive the long journey. You would need tanks of **oxygen** so you could breathe, and specially-designed spacecraft that could both fly and hold up under very strong **pressure.**

From the water ice on Jupiter's **moons,** you could get drinking water, oxygen, and **hydrogen** fuel. However, none of this would help you survive the high levels of **radiation** around Jupiter. You would not last more than a couple of minutes without a heavy-duty spacecraft. Jupiter's strong **magnetic field** holds the radiation close to the planet.

The Space Shuttle Discovery *took the Hubble Space* **Telescope** *into space. Much of the information that scientists have about the planets comes from this telescope.*

You can "visit" with a space probe

Space probes have provided much of the information that we have about Jupiter. There have been five Jupiter missions. The *Pioneer 10* and *11* missions took the first photos of Jupiter and its moons in 1973 and 1974. We also learned more about the magnetic field and the high levels of radiation. *Pioneer 11* discovered Jupiter's **ring** system. In 1979, the *Voyager 1* and *2* space probes showed us details of Jupiter's clouds, moons, and ring system as they flew by Jupiter on their way to study Saturn, Uranus, and Neptune.

The *Galileo* probe

The National Aeronautics and Space Administration (NASA) and the European Space Administration (ESA) launched the *Galileo* space probe on October 18, 1989. The **orbiter** part of the space probe is still flying around Jupiter and sending back information. The orbiter circles Jupiter and studies the planet from outer space. A special **probe** was sent from the spacecraft in July 1995 to collect information about Jupiter's atmosphere. It worked for 57 minutes before the high temperature and pressure destroyed it.

An artist has drawn this picture of the Galileo orbiter receiving data from a probe in Jupiter's clouds. The blue dots are meant to show the data signals.

Fact File

	JUPITER	EARTH
Average distance from the Sun	484 million miles (778 million kilometers)	93 million miles (150 million kilometers)
Revolution around the Sun	11.9 Earth years	1 Earth year (365 days)
Average speed of orbit	8.1 miles/second (13 kilometers/second)	18.6 miles/second 30 kilometers/second
Diameter at equator	88,846 miles (142,984 kilometers)	7,926 miles (12,756 kilometers)
Time for one rotation	9 hours, 50 minutes	24 hours
Atmosphere	hydrogen and helium	nitrogen, oxygen
Moons and rings	at least 16 moons and 3 rings	1 moon and no rings
Temperature range	(atmosphere to core) −235°F (−150°C) to 30,000°F (17,000°C)	(surface only) −92°F (−69°C) to 136°F (58°C)

Color filters were used to make this partial view of Jupiter more dramatic.

Jupiter is shown here with Io, the planet's closest moon, to its left.

A trip to Jupiter from Earth

- When Jupiter and Earth come closest to each other in their **orbits,** they are 367 million miles (591 million kilometers) apart.
- Traveling to Jupiter by car at 70 miles (113 kilometers) per hour, the trip would take at least 600 years.
- Traveling to Jupiter by rocket at 7 miles (11 kilometers) per second, the trip would take at least 607 days.

More interesting facts

- Jupiter spins around on its **axis** about 27 times faster than Earth does. Jupiter is shaped like a slightly flattened ball because it spins so quickly.
- Jupiter has a very tiny **moon** called Leda. It is only about 10 miles (16 kilometers) wide. This means that it is more than 200 times smaller than Earth's moon.
- Metis is Jupiter's fastest-moving moon.

Glossary

ammonia   strong-smelling gas found in Jupiter's atmosphere

asteroid   large piece of floating rock that formed at the same time as the planets and orbits the Sun

astronomer   person who studies objects in outer space

atmosphere   all of the gases that surround an object in outer space

aurora   glowing bands of light high in the atmosphere near the poles of a planet, caused by electrical particles crashing into gases in the atmosphere

axis   imaginary line through the middle of an object in space, around which it spins as it rotates

belt   dark layer of clouds wrapped around Jupiter that alternates with zones

carbon   element found in all plants and animals

chemical   form of matter, or substance

comet   ball of ice and rock that orbits around the Sun

compound   combination of two or more elements

core   material at the center of a planet

crater   bowl-shaped hole in the ground that is made by a meteorite or a burst of lava

crust   top, solid layer of an object in outer space. The outer part of the crust is called the surface.

day   time it takes for a planet to spin around its axis one time

equator   imaginary line around the middle of a planet

gravity   invisible force that pulls an object toward the center of another object in outer space

helium   gas found on many planets; used on Earth to make balloons float in the air

hydrogen   substance found on many planets. On Earth, hydrogen gas mixes with oxygen gas to form water.

magnetic field   region in which the motion of electrical particles and magnets is affected; protects a planet from the wind of electrical particles coming from the Sun

mantle   middle layer of a planet or moon. It lies between the core and the crust.

meteorite   piece of rock or dust that lands on the surface of a planet or moon from space

methane   chemical found in gas form in many atmospheres

molecule   tiny unit made up of two or more atoms of a substance joined together

molten   melted by heat into a liquid form

moon   object that floats in an orbit around a planet

nitrogen   gas found in the atmosphere of Earth and some of the other planets in our solar system

orbit curved path of one object in space moving around another object; or, to take such a path under the influence of gravity

orbiter spaceship that flies in orbit around a planet

oxygen gas that is found in the atmospheres of some planets; used by humans and animals to breathe

pressure force in the atmosphere or inner parts of a planet that presses in from all sides, due to the weight of all material above and around

probe name used for a part of a space probe that leaves the orbiter and studies the atmosphere or surface of an object in space

radiation energy released in waves, particles, or rays. Heat and light are types of radiation.

revolution time it takes for a planet to travel one time around the Sun or for a moon to travel around a planet

ring circle-shaped group of dust and tiny rocks that travels in a close orbit around a planet

solar system group of objects in outer space that all float in orbits around a central star

space probe ship that carries computers and other instruments to study objects in outer space

sulfur yellow-colored, powdery material; found on many planets in gas form

surface part of a planet's crust layer that lies just below the planet's atmosphere

telescope instrument used by astronomers to study objects in outer space

volcano mountain built up from layers of hardened lava

zone light band of clouds wrapped around Jupiter that alternates with belts

More Books to Read

Brimner, Larry Dane. *Jupiter.* Danbury, Conn.: Children's Press, 1999.

Kerrod, Robin. *Jupiter.* Minneapolis, Minn.: Lerner Publications, 2000.

Landau, Elaine. *Jupiter.* Danbury, Conn.: Franklin Watts, 1996.

Index